THE EARTH

TIM FURNISS

HODDER
Wayland

an imprint of Hodder Children's Books

Spinning through space

THE EARTH

Other titles in the series: The Moon ● The Sun ● The Solar System

Find Wayland on the Internet at http://www.wayland.co.uk

All Wayland books encourage children to read and help them improve their literacy.

✓ The contents page, page numbers, headings and index help locate specific pieces of information.

✓ The glossary reinforces alphabetic knowledge and extends vocabulary.

✓ The further information section suggests other books dealing with the same subject.

Cover photographs:
A satellite image of the USA [main]; the *Meteosat* satellite that orbits the Earth [inset top]; a rainbow and waterfall in Iceland [inset middle]; Pinnacle Desert in Western Australia [inset bottom].

Title page: The Earth seen from space by the astronouts of *Apollo 15*.

First published in Great Britain in 1999 by
Wayland (Publishers) Ltd
Reprinted in 2000 by Hodder Wayland,
an imprint of Hodder Children's Books
© Hodder Wayland 1999

Editor: Carron Brown
Designer: Tim Mayer
Production controller: Carol Titchener
Illustrator: Peter Bull

British Library Cataloguing in Publication Data
Furniss, Tim
    The Earth. – (Spinning through Space)
    1. Astronomy – Juvenile literature
    2. Earth – Miscellanea – Juvenile literature
    I. Title
    525

ISBN 0 7502 2409 6

Printed and bound in Italy by EuroGrafica, Vicenza

CONTENTS

THE EARTH IN SPACE

▼ Astronauts on the way to the Moon took this photo of the Earth in 1972.

For us, our planet Earth seems to be the centre of everything. However, the Earth is just one very small part of a vast universe that remains our greatest mystery. For thousands of years, people have gazed into the night sky in sheer wonder. Four thousand years ago, Egyptian astronomers observed the Sun. They made sundials to tell the time. Other astronomers mapped the stars and named the constellations. People were slowly beginning to understand the wonderful universe.

▲ Around 1500 BC, people observed the position of the Sun so that they could build a stone circle that would receive its full light at midsummer. The result is Stonehenge, in the UK.

Passages from the Bible written in about 2000 BC said that the world was round and was suspended in space.

Early astronomers thought that the Earth was at the centre of the universe. People also thought the Earth was flat, and, if you walked too far, you'd fall off the edge! In about 550 BC, the Greek astronomer Pythagoras suggested that the Earth was round. It wasn't until 1543 that a Polish astronomer, Nicolaus Copernicus, said that the Earth orbited the Sun, along with a number of other planets. Today, spacecraft from the Earth have explored all but one of these planets.

Pluto

Neptune

Uranus

Jupiter

Earth

Mars

Venus

Saturn

The Sun

Mercury

The solar system

We live on a 'spaceship' called Earth. Our planet is travelling through space at a speed of almost 30 km per second.

Nine planets, including the Earth, orbit the Sun in a solar system. Jupiter, the largest planet, is about 318 times more massive than the Earth. Pluto, the smallest planet, is smaller than our Moon. The Earth has just one moon but many planets have a lot of moons. Astronomers have spotted at least 18 moons around Saturn.

The Milky Way

We can see 2,500 stars in the night sky. The Sun is also a star but it shines so brightly that we can't see any other stars during the day. The Sun is on the edge of a galaxy of 100,000 million stars that we call the Milky Way. The Milky Way is only one of millions of galaxies in the universe.

It is possible to see Mercury, Venus, Mars, Jupiter and Saturn by just looking up into the night sky at certain times of the year.

The Sun is 150 million km away from the Earth. It takes sunlight 8 minutes and 17 seconds to reach us. Light from the nearest galaxy to the Milky Way takes 2,200,000 years to reach the Earth. Light from the furthest reaches of the universe that we can see takes 14 thousand million years! So the Earth is a very small part of the universe.

▼ There are millions of other galaxies in the universe. These are the Draco (left) and Sombrero (right) galaxies.

EARTH FACTS

The Earth is 12,756 km in diameter, one of the medium-sized planets in the solar system.

A planet of water

From space, the Earth is a beautiful, bright, bluish planet. The vast amount of water on the Earth reflects the Sun's light like a mirror. The Earth has all three forms of water: ice, liquid and vapour. Water vapour rising from the sea forms huge white clouds and is a huge influence on the weather. The Earth experiences very different types of weather, from hurricanes to droughts.

An island in Hawaii, called Kauai, has recorded over 350 rainy days a year.

▼ Three-quarters of the Earth is covered by seas and oceans.

▲ A burning hot desert in Saudi Arabia.

▲ The Himalayas as seen from space. Mount Everest is part of this mountain range.

The Pacific Ocean represents about 45 per cent of the world's seas.

The varied surface

The Earth has vast areas of land, varying from lush vegetation to harsh hot and cold sandy deserts. It has high mountain ranges and large areas of flat plains. At the North and South Poles of the Earth, there are huge masses of ice called the polar caps. The deepest point on the Earth is the Marianas Trench. This is 11,034 m below the Pacific Ocean. The highest point on the Earth, at 8,848 m high, is Mount Everest, on the border between Nepal and Tibet.

Temperatures

Temperatures on the Earth vary from about minus 89°C to about 49°C. Most parts of the Earth experience less extreme temperatures which make it possible to live comfortably.

This artwork shows ▶
the layers of the Earth.

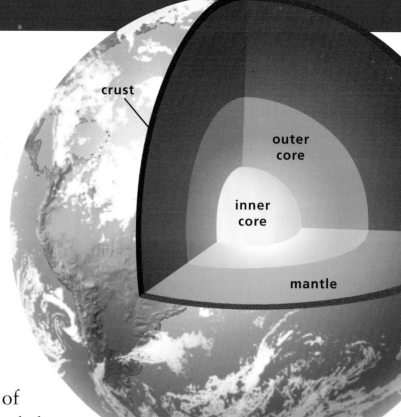

crust

outer
core

inner
core

mantle

Below the surface

The Earth has a core of molten iron, 4,800 km below the surface. It has a crust about 32 km deep. Molten lava sometimes oozes through cracks in the thin mantle, forming volcanoes. Movements in the crust cause earthquakes.

The atmosphere

The Earth is surrounded by a very thin layer of air, called the atmosphere. It is made up of various gases that are vital to support life on the Earth. The two main gases, nitrogen and oxygen, allow us to breathe. Plants need the gas carbon dioxide to grow. Bacteria in the soil need nitrogen to make food for plants and animals.

In 1798, scientist Henry Cavendish correctly calculated the weight of the Earth to be about 6,000,000,000,000,000,000,000 tonnes!

Of 10,000 molecules in the air, 7,809 are nitrogen, 2,095 are oxygen, 93 are argon and 3 are carbon dioxide.

▼ The Earth's atmosphere seen from space.

At lower levels, the atmosphere also contains water vapour. When this cools, it can form droplets of rain or flakes of snow. Most of the atmosphere is held in place by the Earth's gravity. The lower the air, the more dense the atmosphere is. About nine-tenths of the atmosphere is packed into the first 16 km above the surface.

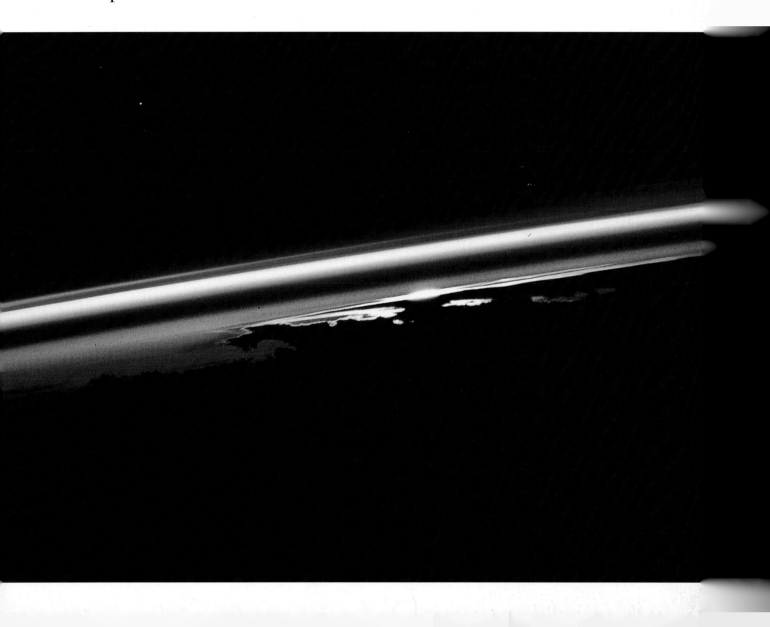

◀ A rainbow shows the different colours of sunlight.

We can see more than a hundred shades of the six colours of the rainbow.

The spectrum

We call the light that comes from the Sun, white light. However, sunlight is made up of six colours, called a spectrum. There are six main colours which we can see in a rainbow: red, orange, yellow, green, blue and violet. A rainbow is formed when the Sun's light is divided into its separate colours by raindrops. The colours grow stronger when the raindrops are large.

Four hundred years ago, the famous scientist Sir Isaac Newton was the first person to separate sunlight into its separate colours.

Why is the sky blue?

Light from the Sun hits the molecules of the air in the Earth's atmosphere. It is scattered in all directions. We see blue light in the sky because is scattered more than other colours. The sky turns red at sunset because the Sun's light has to travel through more atmosphere when it is low in the sky. The blue light is stopped but more red light gets through.

▼ This illustration shows how blue light is scattered when it hits the air in the atmosphere.

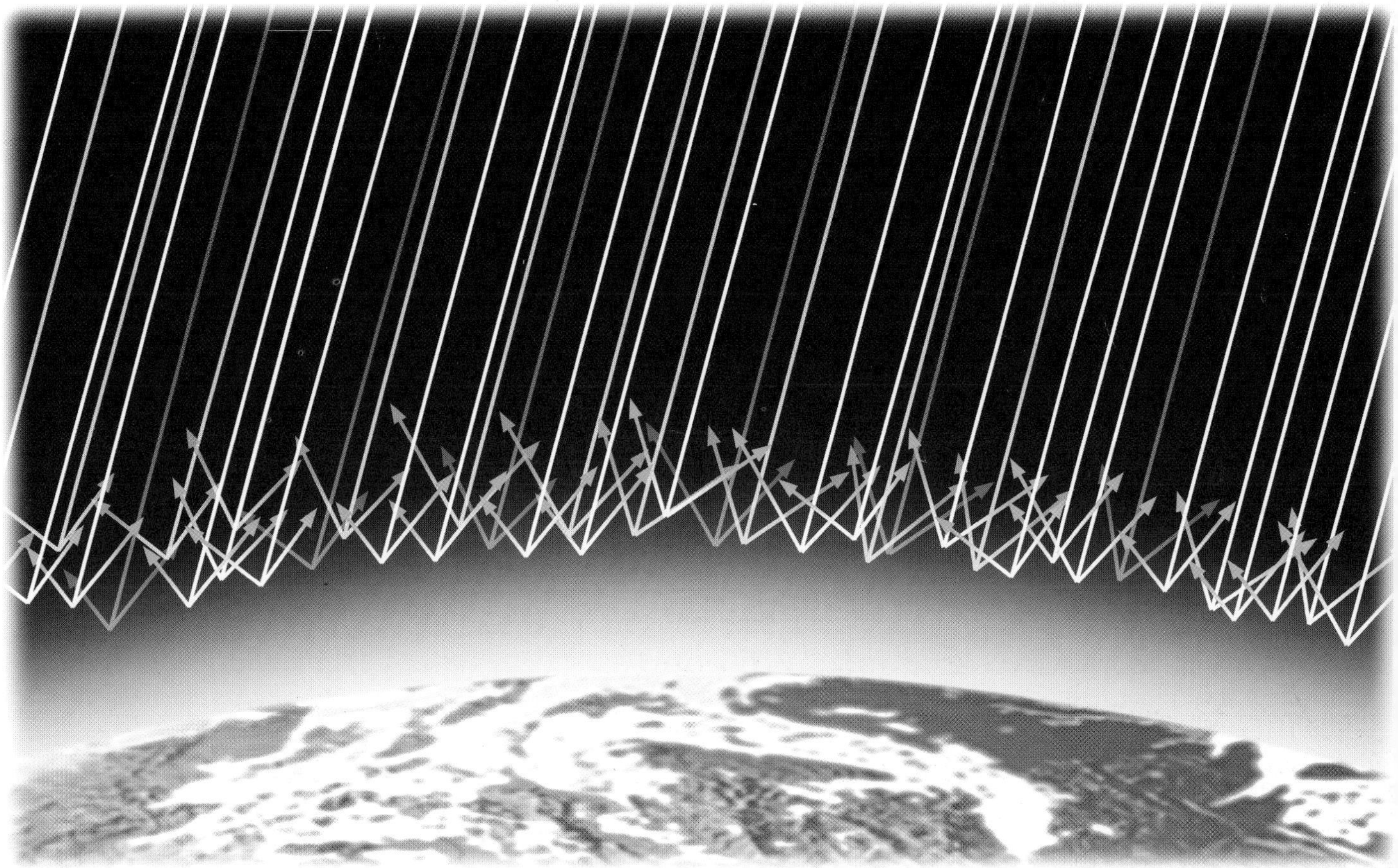

The ozone layer

The atmosphere shields us from meteoroids. When these hit the atmosphere, they are travelling so fast that friction against the atmosphere causes them to heat up. They turn into meteors or 'shooting stars'.

The atmosphere has a layer of ozone gas that helps to cut out some of the harmful radiation that comes from the Sun. However, this protective shield has been damaged by harmful chemicals used on Earth, causing holes to appear. Since scientists discovered the damage, the chemicals are not being used any more and the ozone layer is slowly recovering.

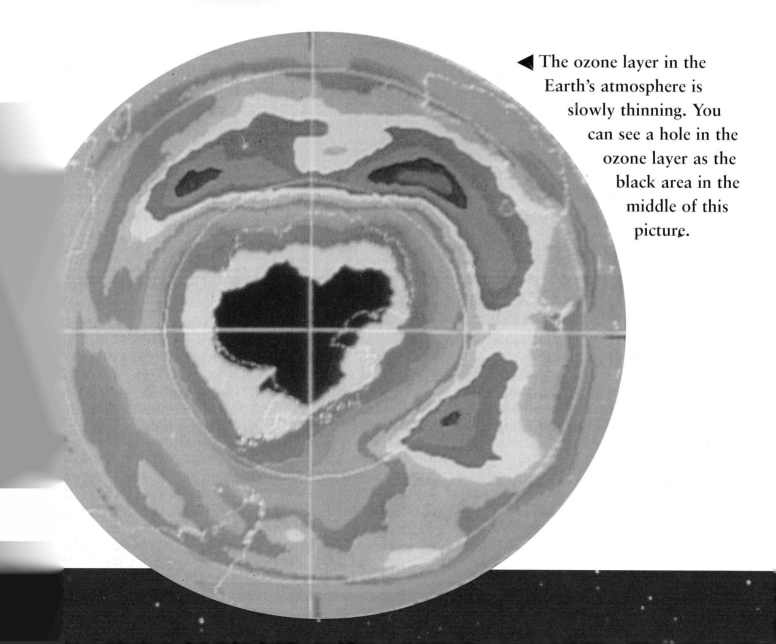

◀ The ozone layer in the Earth's atmosphere is slowly thinning. You can see a hole in the ozone layer as the black area in the middle of this picture.

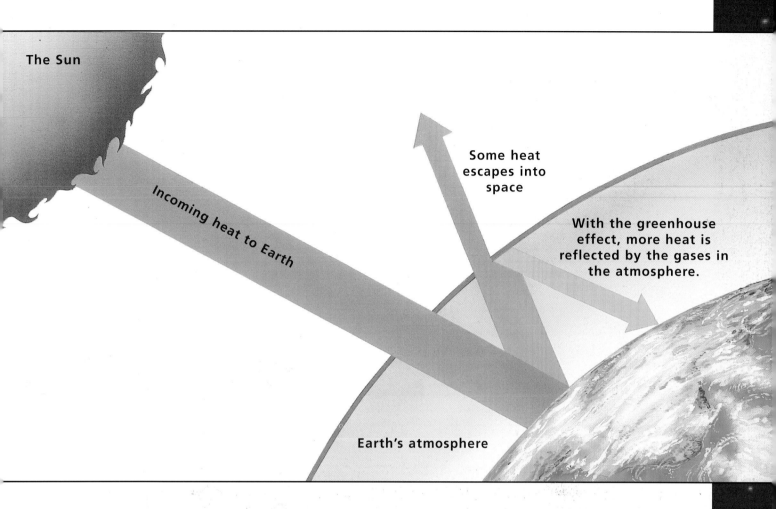

The Sun

Incoming heat to Earth

Some heat escapes into space

With the greenhouse effect, more heat is reflected by the gases in the atmosphere.

Earth's atmosphere

The greenhouse effect

Pollution is also causing carbon dioxide to increase the size of the atmosphere, trapping the Sun's heat on the Earth's surface. It is a bit like being inside a giant greenhouse, where the Sun's heat is trapped and cannot escape. Scientists are trying to work out how to help our planet to cope with the extra heat on the planet's surface.

▲ This illustration shows how the heat is trapped on the Earth's surface.

Without the layers of gases in the atmosphere, the Earth would be as cold and lifeless as the Moon.

LIFE ON EARTH

Rodents, such as rats, are the most common form of mammal in the world. They account for 50 per cent of the species of living mammals.

▼ There are about 5,000 million people on the Earth.

Earth is unique. It is the only planet, and the only place in the universe, where we know that life definitely exists. It is a life of incredible variety: such as mammals, birds, reptiles, insects, fish and plants.

There are about 275,000 different types of flowering plants and trees. Over one million insects have been discovered. There are over 500,000 species of animal. The Earth is just the right distance from the Sun to allow life to develop. The Earth has a perfect atmosphere to support life and has a lot of fresh water in liquid form.

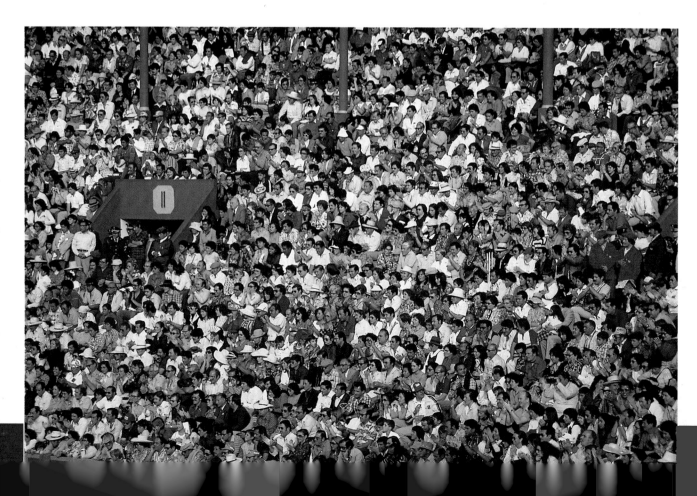

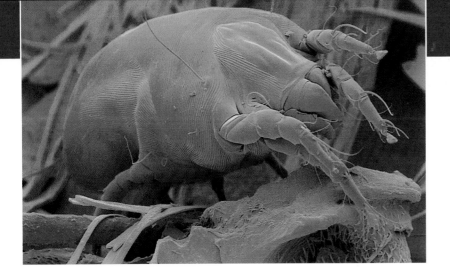

◄ The dust mite is one of the smaller living creatures.

The Earth's seasons cause a variety of weather conditions which have helped different life forms to develop. Most scientists believe that this extraordinary range of life just happened to start 3,500 million years ago, as simple microscopic living cells. These formed in water.

Scientists say that water is the most important requirement for life to form. They have also found evidence to show that humans were on the Earth about two million years ago.

▼ Blue whales can be 27 m long and weigh 120 tonnes.

GRAVITY

Why do we stay on the ground and not float over the Earth? We stay on the ground because of gravity, the force that is pulling towards the Earth's centre. That direction is down. When you hold a stone, you can feel its weight because of the gravity pulling it down into your hand.

Sir Isaac Newton was the first person to understand how gravity works. He realized what gravity was when he sat under an apple tree and an apple fell down on to his head. Gravity works between two objects, pulling the less massive object to the more massive one. The Earth is the most massive object near to us, so it pulls everything towards it.

A man weighing 65 kg on Earth would weigh 1,800 kg on Saturn. The more massive an object is, the greater the gravity between it and another object. Saturn is very massive.

▼ These arrows show some of the forces that work on the Earth.

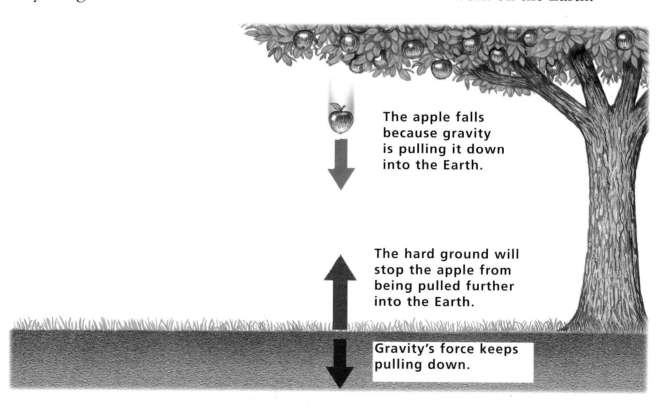

The apple falls because gravity is pulling it down into the Earth.

The hard ground will stop the apple from being pulled further into the Earth.

Gravity's force keeps pulling down.

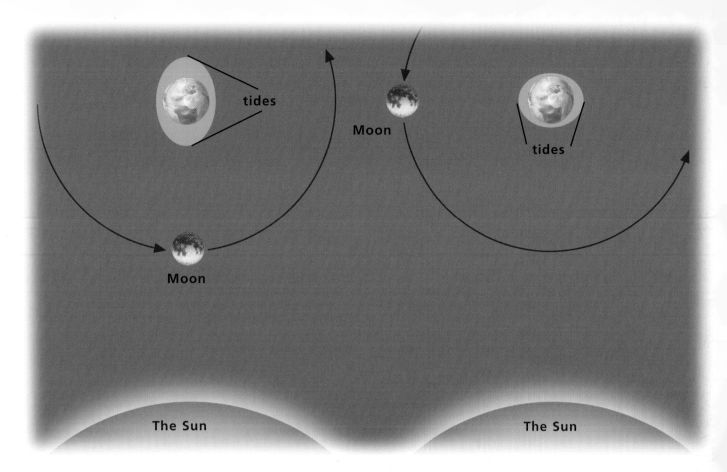

tides

Moon

Moon

tides

The Sun

The Sun

▲ The Moon's pull of gravity causes tides as it orbits the Earth. The tides follow the Moon's movement. Sometimes, the Sun adds to the pull, creating larger tides.

In that case, why doesn't the Moon fall on us? The Moon does not drop on the Earth because it is moving very fast. However, it doesn't escape the pull of gravity altogether, so it orbits the Earth. A satellite that is orbiting the Earth is actually 'falling' towards the planet but is travelling so fast, it falls 'around' the Earth in a continuous orbit like the Moon.

If you can jump 1 m into the air on the Earth, you could jump 4.5 m above the Moon.

The strength of a planet's pull of gravity depends on how massive it is. In our solar system, Jupiter, the largest planet, has a greater pull of gravity than the other planets and Pluto has the least pull.

SPACESHIP EARTH

The Earth is like a spaceship. It orbits the Sun in 365 days, which we call a year. The Earth orbits the Sun at a speed of 105,000 km per hour. The Earth is held in its orbit by the gravity of the Sun and by its own speed. It travels a distance of over 900 million km in a year.

The Earth's orbit around the Sun is not perfectly circular. In July, the Earth is at its closest to the Sun, at a distance of 147 million km. In January, the Sun is at its furthest distance at 152 million km.

The Earth actually rotates every 23 hours 56 minutes and 4 seconds. It orbits the Sun every 365 days 6 hours 9 minutes and 10 seconds.

▼ The Sun's gravity keeps the planets in orbit around it.

The planets:
1. Mercury
2. Venus
3. Earth
4. Mars
5. Jupiter
6. Saturn
7. Uranus
8. Neptune
9. Pluto

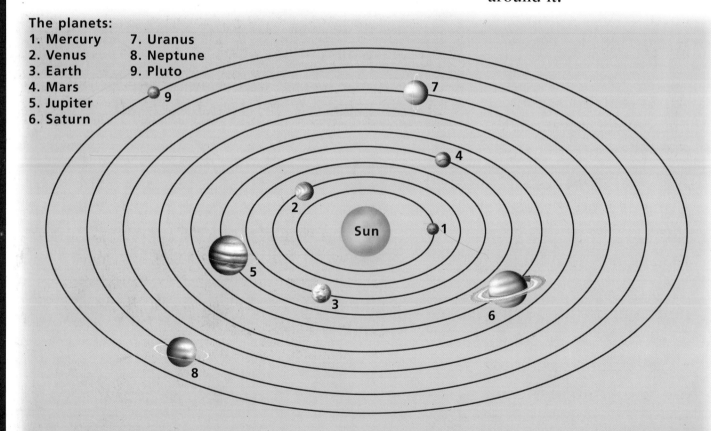

When one side of the ▶ Earth faces the Sun, the other side is dark.

While the Earth orbits the Sun, it also rotates. One rotation of the Earth takes 24 hours. We call this is a day.

As the Earth rotates, the Sun seems to rise in the east and set in the west. At night, we can see the stars going around the sky very slowly. This is also because the Earth is spinning round. We can see some of the other planets in the sky, too. They very slowly change position. This is because they are also orbiting the Sun. The planets all move at different speeds.

Every four years, we have one extra day, on 29 February. This is called a leap year. The extra day is needed because of the extra 6 hours, 9 minutes and 10 seconds in the Earth's orbit around the Sun. Over 4 years, these add up to about 24 hours.

THE SEASONS

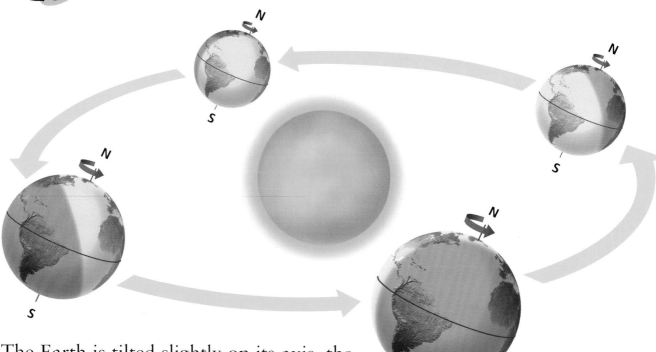

The Earth is tilted slightly on its axis, the imaginary line running through the Earth's poles. The axis is tilted 23.5 degrees. If you keep looking to the north on a dark night, you will see the stars seem to go around one point in the sky. The Earth's North Pole points to this one point, called the Pole Star.

As the Earth orbits the Sun, the North Pole sometimes tilts more to the Sun than the South Pole. Later in the year it is the other way round. This means that the northern or southern hemispheres have different lengths of days. When the northern hemisphere has long days and short nights, the southern hemisphere as short days and long nights. This creates the seasons.

▲ This illustration shows how the Earth's tilt causes the seasons as it orbits the Sun.

If the Earth's axis pointed straight up, the areas at North and South Poles would have a year of permanent sunlight and a year of permanent darkness.

Seasons are times of the year when the Sun is in the sky for different lengths of time. In the summer, the Sun rises early and sets late. In the winter, it is the opposite. The seasons in between are called spring and autumn. When it is winter in the northern hemisphere, it is summer in the southern hemisphere.

The tilt of the Earth's axis sometimes causes the polar regions to have 24 hours of Sun or 24 hours of darkness. When the North Pole leans towards the Sun in June, the extreme north has 24 hours of daylight while the extreme South has 24 hours of night. In December, this is reversed.

In midsummer, the Sun ▶ is high in the sky, while in midwinter it is very low.

THE SPACE CLOCK

Time is based on the Earth's movement. One rotation of the Earth takes 24 hours, which gives us our first measurement, a day. Our clock is based on the position of the Sun as the Earth rotates. When the Sun is at its highest in the sky, it is 12 hours into the day. This is called midday. Twelve hours later, it is midnight.

The world's timekeeping starts at Greenwich in London. An imaginary line called the Prime Meridian, or 0° longitude, runs from the North to the South Pole through Greenwich. The time at Greenwich is called Greenwich Mean Time (GMT) UK. Time at other parts of the world is measured in relation to GMT.

Sosigenes, a Greek astronomer who worked for the Roman leader Julius Caesar, devised the monthly calendar. July is named after Julius. August is named after Augustus Caesar, the next Roman ruler. The calendar is based on the phases of the Moon – there are about 28 days betweeen one 'new Moon' and the next.

▼ The time zones.

| 24 | 1 | 2 | 3 | 4 | 5 | 6 | 7 | 8 | 9 | 10 | 11 | 12 noon | 13 | 14 | 15 | 16 | 17 | 18 | 19 | 20 | 21 | 22 | 23 | 24 |

GMT

GMT

Standard times with alternate hours from Greenwich.

Area where time varies from standard time by half an hour or more.

Time is divided into 12 time zones around the world. To work out the time in the zones going east of GMT, add one hour to each time zone from the one before. For the zones to the west of GMT, take away one hour from the zone before.

So when it is 12 o'clock midday GMT, it is 11 a.m. in the next zone going west across the Atlantic Ocean, but it is 1 p.m. in Oslo, Norway, in the next zone towards the east.

▲ When it's 9 p.m. in Tokyo (above), it's midday in London (below). So Tokyo is nine time zones east of GMT. ▼

Sundials work by using the position of the Sun in the sky. The Sun casts a shadow on to the clock-face on a sundial. But the Sun has to be shining for the sundail to work.

THE EARTH FROM SPACE

No one knew what the Earth looked like from space until *Apollo 8* blasted into orbit in 1968. Its mission was to orbit the Moon for the first time. One of the astronauts was James Lovell. When he saw the whole globe of the Earth from space, he described it as a 'fragile oasis in the vastness of space'. He realized that the Earth is a very small but very special part of the vast universe.

Twenty-two *Apollo* astronauts have seen the Earth rise above the Moon.

◀ The *Apollo 8* crew, left to right: Frank Borman, Bill Anders and James Lovell.

▲ The most exciting sight for the *Apollo* astronauts: Earth rising over the Moon.

If the Earth was as big as Jupiter, it would fill the whole earthrise image.

One of the major benefits of space research has been to give us this view of the Earth. Scientists dream of flights to the planets and to the stars, but the sight of the beautiful planet Earth in space draws us home.

Lovell had two companions on *Apollo 8*, Frank Borman and Bill Anders. These three astronauts were the first to be cut off from the sight of Earth as they flew around the Moon. Later, they saw the Earth rising like the Sun. The picture of the earthrise that they took has become one of the most famous to be taken in space.

OBSERVING EARTH

We have learned a tremendous amount about the Earth and its environment from satellites. Weather satellites take daily images of the Earth's weather, its atmosphere and its temperature. Other satellites survey its resources and monitor pollution.

A number of satellites are orbiting around Earth's equator at a speed of 36,000 km per hour. This is the same speed as the Earth rotates. This is called a geostationary orbit. So, the satellites always seem to be in the same place overhead.

Weather satellites in geostationary orbit show the Earth's weather, the temperature of the land and sea, the amount of water vapour circulating in the atmosphere and lots of other information all at once.

▼ This is a *Meteosat* weather satellite orbiting the Earth.

This satellite orbits the ▶
Earth and takes radar
images of its surface.

These satellites are stationed over different areas
of the Earth. So a network of them covers the
whole globe. They send back pictures every day.
Other satellites in lower orbits are used
to take close-up pictures of certain
parts of the Earth. Most of them
orbit around the poles of the Earth.
So, in one day, they go around the
Earth about 17 times. As they are
going around, the Earth is rotating
beneath. The satellites can cover the whole
globe in a day.

The first pictures of the world's
weather from space were taken in 1960
by the *Tiros 1* satellite.

GLOSSARY

Astronauts People who travel through space in a spacecraft.

Astronomers People who study space.

Axis The imaginary line through the Earth from the North to the South Poles.

Bacteria Very tiny living things.

Cells Very tiny parts of a living body, such as blood cells.

Constellations Groups of stars in the night sky. They may not be in the same galaxy as ours.

Dense Thick.

Droughts Dry weather that lasts for a long time.

Equator The imaginary line that runs round the centre of the Earth, halfway between the poles.

Galaxy A group of millions or billions of stars in the sky.

Gravity A force between two objects. Smaller objects are drawn to bigger objects.

Globe Anything that is ball-shaped, spherical.

Hemispheres Halves of spheres. The Earth is split up into two hemispheres, called the northern and southern hemispheres.

Hurricanes Violent storms with very strong winds.

Mass The amount of substance in an object.

Meteoroids Solid objects floating in space.

Molecules The smallest parts of a substance.

Orbit To go round.

Phases The different shapes of the Moon that we can see as it orbits the Earth, caused by light from the Sun.

Planets Objects in space that orbit stars.

Population The people living in a place.

Rotates To turn around on a centre or axis.

Solar system The Sun, its nine planets and moons. 'Solar' means to do with the Sun.

Stars Large luminous points in space that are spherical and made up of many different gases.

Sundials Clocks that allow people to tell the time by looking at the position of shadows falling on a clock-face.

Trillion A million million.

Universe Everything that is in space.

Vapour Tiny invisible droplets of water in the air.

FURTHER INFORMATION

Web pages:

www.fourmilab.ch/earthview/vplanet.html   Earth and Moon viewer

earth.jsc.nasa.gov/   Earth photos

Books to read:

A Closer Look at The Ozone Hole by Alex Edmonds (Watts, 1996)

Earth, Gateway (*Solar System* series) by Gregory Vogt (Amazon, 1996)

Earth and Space (*Usborne Starting Point Science* series) by Susan Mayes and Sophy Tahta (Usborne, 1995)

The Earth in Space (*Straightforward Science* series) by Peter Riley (Watts, 1998)

The Kingfisher Book of Space by Martin Redfern (Kingfisher, 1998)

The Planets by Patrick Moore (Aladdin, 1994)

Stars and Planets by David H. Levy (Macdonald Young, 1996)

Places to visit:

The Science Museum, Exhibition Road, South Kensington, London (Tel: 0171 938 8000) has many exhibits about rockets, satellites and other spacecraft.

The Planetarium, Euston Road, London (Tel: 0171 935 6861) has programmes about planets, space and the stars, and how they are explored by spacecraft and satellites.

THE EARTH

HISTORY
- Find out how the months got their names: i.e., famous astronomers.
- Look at how people thought of the Earth through the ages: i.e., the world being flat; discovery that it's spherical.

MUSIC
- Compose a piece of music that describes the different types of weather on Earth.

MATHS
- Measurement: e.g., diameter, size, weight, distance from the Sun.
- Explain that planets are spherical.

DESIGN AND TECHNOLOGY
- Look at how buildings are built in order to withstand earthquakes.

ENGLISH
- Write a newspaper article about why we need to protect our atmosphere.
- Read myths and legends about the Earth, the weather, Sun and Moon.

ART AND CRAFT
- Draw your own map of the solar system with the planets circling the Sun in the correct order.
- Draw your own earthrise picture using the photograph on page 27.

SCIENCE
- Look at how carbon dioxide in our atmosphere is used by plants.
- Investigate the millions of different types of insects to get an idea of the incredible variety of life on Earth.
- Look at how the Earth's orbit around the Sun creates a year; how its rotation creates a day; look at the different time zones.
- Explain how gravity works as a force between two objects, and why the planets orbit the Sun.

GEOGRAPHY
- Compare temperatures all overf the world.
- Find Mount Everest and the Marianas Trench in an atlas.
- Investigate how movements in the crust of the Earth cause earthquakes and volcanic eruptions.

INDEX

All numbers in **bold** refer to pictures as well as text.

Picture aknowledgements:

The publishers would like to thank the following for allowing us to reproduce their pictures in this book: Eye Ubiquitous *cover* [inset middle], 23, /Paul Thompson 16; Genesis *cover* [inset top], *title page*, *imprints page*, 4, 6, 9 [bottom], 14, 21, 26, 27, 28, 29; Robert Harding 25 [top], /C. Bowman *cover* [inset bottom], /Flip Nicklin 17 [bottom], /Amanda Tovy 25 [bottom]; Science Photo Library/Ray Ellis 9 [top], /Esa/Photo Library International *cover* [main], /Michael Morven 12, /NASA *contents page*, 8, 11, /David Parker 5; Andrew Syred 17 [top].

The illustrations on pages 6, 10, 13, 15, 18, 19, 20, 22 and 24 are by Peter Bull.